U0315071

2015

室内设计模型集成
INTERIOR DESIGN MODELS
INTEGRATION

国广一叶装饰机构
叶斌 叶猛 著

公共空间 PUBLIC SPACES

海峡出版发行集团 | 福建科学技术出版社
THE STRAITS PUBLISHING & DISTRIBUTING GROUP | FUJIAN SCIENCE & TECHNOLOGY PUBLISHING HOUSE

叶 斌／
Ye Bin

高级建筑师
国广一叶装饰机构首席设计师
福建农林大学兼职教授
南京工业大学建筑系建筑学学士
北京大学 EMBA
中国建筑学会室内设计分会理事
中国建筑装饰协会理事

Senior Architect
Chief Architect of Guoguangyiye Decoration Group
Adjunct Professor of Fujian Agriculture and Forestry University
B. Arch from Nanjing Industry University
EMBA from Beijing University
Councilor Member of China Institute of Interior Design
Councilor Member of China Building Decoration Association

著作

1. 《室内设计图典》（1、2、3）
2. 《装饰设计空间艺术·家居装饰》（1、2、3）
3. 《装饰设计空间艺术·公共建筑装饰》
4. 《建筑外观 Appearance 细部图典》
5. 《国广一叶室内设计模型库·家居装饰》（1、2、3）
6. 《国广一叶室内设计模型库·公建装饰》
7. 《国广一叶室内设计》
8. 《国广一叶室内设计模型库构成元素》（1、2）
9. 《室内设计立面构图艺术》系列
10. 《国广一叶室内设计模型库》系列
11. 《家居装饰·平面设计概念集成》
12. 《概念家居》
13. 《概念空间》
14. 《国广一叶家居装饰》系列
15. 《室内设计图像模型》系列
16. 《细解家居》系列
17. 《2009 室内设计模型》系列（5 册）
18. 《2010 家居空间模型》系列（3 册）
19. 《2010 公共空间模型》系列（2 册）
20. 《2011 家居空间模型》系列（3 册）
21. 《2011 公共空间模型》
22. 《最给力家装》系列（6 册）
23. 《2012 室内设计模型集成》系列（5 册）
24. 《名家家装 + 材料标注》系列（5 册）
25. 《2013 家居空间模型集成》系列（3 册）
26. 《2013 公共空间模型集成》系列（2 册）
27. 《2014 家居空间模型集成》
28. 《2014 公共空间模型集成》

荣誉

荣获"中国室内设计杰出成就奖"
当选 2009 年"金羊奖"中国十大室内设计师
当选中国建筑装饰行业新中国成立 60 年百名功勋人物
当选 1989~2009 年中国杰出室内设计师
当选 1997~2007 年中国家装十年最具影响力精英领袖
当选 1989~2004 年全国百位优秀室内建筑师
当选 2004 年度中国杰出中青年设计师
当选 2004 年度中国室内设计师十大封面人物
当选 2002 年福建省室内设计十大影响人物（第一席位）

获奖设计作品

溪山温泉度假酒店（实例）	2014 年第十届中国国际室内设计双年展金奖
正兴养老社区体验中心	2014 年第十届中国国际室内设计双年展银奖
中联大厦办公楼	2014 年第十届中国国际室内设计双年展铜奖
尊贵彰显富丽	2014 年第十届中国国际室内设计双年展铜奖
永福设计研发中心	2014 年度全国建筑工程装饰奖（公共建筑装饰设计类）
宇洋中央金座	2013 年第十六届中国室内设计大奖赛铜奖
书·韵	2013 年国际空间设计大奖"Idea-Tops 艾特奖"提名奖
聚春园驿馆	2013 年国际空间设计大奖"Idea-Tops 艾特奖"入围奖
童话地中海	2013 年国际空间设计大奖"Idea-Tops 艾特奖"入围奖
宇洋中央金座	2013 年国际空间设计大奖"Idea-Tops 艾特奖"入围奖
宁德上东曼哈顿销售楼部	2013 年第四届中国国际空间环境艺术设计大赛（筑巢奖）优秀奖
福建洲际国际酒店	2012 年第二届亚太酒店设计大赛金奖
前线共和广告	2012 年第十五届中国室内设计大奖赛金奖
瑞莱春堂福州三坊七巷店	2012 年"照明周刊杯"中国照明应用设计大赛一等奖
前线共和广告	2012 年第九届中国国际室内设计双年展金奖
福州情·聚春园	2011 年第九届中国国际室内设计双年展金奖
宁化世界客家文化交流中心	2011 年第九届中国国际室内设计双年展铜奖
一信（福建）投资	2011 年第十四届中国室内设计大奖赛金奖
福建科大永合医疗机构	2011 年中国最成功设计大赛最成功设计奖
连江贵安海峡文化村酒店	2011 年中国（上海）设计节"金外滩"最佳概念设计奖
素丽娅泰水疗会所	2010 年第八届中国室内设计双年展金奖
摩卡小镇售楼中心	2010 年第八届中国室内设计双年展银奖
大洋鹭洲	2010 年第八届中国室内设计双年展铜奖
素丽娅泰水疗会所	2010 年亚太室内设计双年大奖赛商业空间设计银奖
中联江滨御景会所	2010 年亚太室内设计双年大奖赛商业空间设计优秀奖
繁都魅影	2010 年亚太室内设计双年展大奖赛住宅空间设计银奖
繁都魅影	2010 年亚太室内设计大奖赛铜奖
中央美苑	2010 年海峡两岸室内设计大赛金奖
繁都魅影	2010 年海峡两岸室内设计大赛金奖
光·盒中盒	2010 年海峡两岸室内设计大赛金奖
中联江滨御景会所	2010 年海峡两岸室内设计大赛银奖
皇帝洞书院	2009 年"尚高杯"中国室内设计大奖赛二等奖（全国商业类第三名）
北湖皇帝洞景区会所	2008 年第七届中国室内设计双年展金奖
国广一叶点房财富中心	2007 年福建省室内设计大奖赛一等奖（公建工程类第一名）
国广一叶大会会馆	2006 年福建省室内设计大奖赛一等奖（公建工程类第一名）
点房财富中心	2007 年"华耐杯"中国室内设计大奖赛二等奖（全国商业类第二名）
大家会馆	2006 年第六届中国室内设计双年展金奖
书香大第销售中心	2006 年第六届中国室内设计双年展金奖
金钻世家某单元房	2006 年第六届中国室内设计双年展银奖
福州金龙门餐厅	2006 年第六届中国室内设计双年展银奖
滨江丽景·美丽园	2006 年第六届中国室内设计双年展银奖
福建电力调度通信中心大楼	2006 年第六届中国室内设计双年展铜奖
金源国际酒店桑拿中心	2006 年第六届中国室内设计双年展优秀奖
内蒙古呼和浩特市中级人民法院	2004 年第五届中国室内设计双年展铜奖
厦门奥林匹亚中心	2004 年第五届中国室内设计双年展铜奖
福州玖玖丰田汽车 4S 店	2004 年第五届中国室内设计双年展优秀奖

叶 猛／
／Ye Meng

国广一叶装饰机构副总设计师
国家一级注册建筑师
国家一级注册建造师
中国建筑学会室内设计分会会员
福建工程学院建筑与规划系讲师
福州大学建筑系学士
中南大学土建学院建筑学硕士

Deputy Chief Architect of Guoguangyiye Decoration Group
First-Class Registered Architect (PRC)
Registered Constructor (PRC)
Member of Institute of Interior Design of Architectural Society of China
Lecturer of Architecture and Planning Dept., Fujian University of Technology
B. Arch from Fuzhou University
M. Arch from Central South University

获奖设计作品

融信大卫城	2014 年第十届中国国际室内设计双年展优秀奖
三盛国际公园	2014 年第五届中国国际空间环境艺术设计大赛（筑巢奖）提名奖
名城港湾	2014 年第五届中国国际空间环境艺术设计大赛（筑巢奖）优秀创意奖
融侨外滩	2014 年第五届中国国际空间环境艺术设计大赛（筑巢奖）优秀创意奖
鳌峰洲小区—19A	2013 年第四届中国国际空间环境艺术设计大赛（筑巢奖）优秀奖
阳光理想城	2011 年第九届中国国际室内设计双年展金奖
大洋鹭洲	2010 年第八届中国室内设计双年展铜奖
繁都魅影	2010 年亚太室内设计大奖赛铜奖
福建工程学院建筑系新馆	2009 年中国室内空间环境艺术设计大赛一等奖
福建工程学院建筑系新馆	2009 年福建室内与环境设计大奖赛公建工程类最高奖
文化主题酒店	2008 年福建省第六届室内与环境设计大赛一等奖
旗山文城	2008 年福建省第六届室内与环境设计大赛一等奖
另类博弈	2008 年第六届现代装饰年度办公空间大奖入围奖
点房财富中心	2007 年"华耐杯"中国室内设计大奖赛二等奖
翻阅古朴	2007 年福建省第五届室内设计与环境大赛一等奖
大家会馆	2006 年第六届中国室内设计双年展金奖
福建电力调度通信中心大楼	2006 年第六届中国室内设计双年展铜奖

另出版《建筑外观细部图典》、《室内设计图像模型》等著作数十种

国广一叶装饰机构，作为"2012 年度中国建筑装饰设计机构 50 强企业"（中国建筑装饰协会颁发）、"2012～2013 年度全国室内装饰优秀设计机构"（中国室内装饰协会颁发）、"2012 年中国十大品牌酒店设计机构"（中外酒店论证颁发）、"2011～2012 年度全国室内装饰优秀设计机构"（中国室内装饰协会颁发）、"2013 中国住宅装饰装修行业最佳设计机构"（中国建筑装饰协会颁发）、"1989～2009 年全国十大室内设计企业"（中国建筑协会室内设计分会颁发）、"1988～2008 年中国室内设计十佳设计机构"（中国室内装饰协会颁发）、"1997～2007 年中国十大家装企业"（中国建筑装饰协会颁发）、"福建省著名商标"、"省、市级重合同守信用企业"，荣获国际、国家及省市级设计大奖上千项。

国广一叶装饰机构首席设计师叶斌荣获"中国室内设计杰出成就奖"、两次荣获"中国十大室内设计师"称号；叶猛被评为"1989～2009 年中国优秀设计师"；另外，十余位设计师被评为中国装饰设计行业优秀设计师，79 名设计师分别被评为福建省优秀设计师、福州市优秀设计师，56 名在职设计师分别荣获历届全国、福建省、福州市室内设计一等奖……以上这些荣誉的获得和国广一叶装饰机构自身的水准有关。国广一叶装饰机构拥有大批量高水准的室内设计专业效果图，这些效果图将设计师的设计意图淋漓尽致地表现出来。

自 2004 年至今，国广一叶装饰机构在福建科学技术出版社已陆续出版了 12 套模型系列图书，一直受到广大读者的支持与厚爱。为了不辜负广大读者的期望，我们继续推出《2015 室内设计模型集成》系列图书，汇集国广一叶装饰机构 2014～2015 年设计制作的 1600 多个风格各异、手法时尚的室内设计效果图及其对应的 3ds Max 场景模型文件，可作为读者做室内设计时的有益参考。

本书配套光盘的内容包含效果图原始 3ds Max 模型和使用到的所有贴图文件。由于 3ds Max 软件不断升级，此次的模型我们采用 3ds Max2011 版本制作。模型按图片顺序编排，易于查阅调用。只有能对模型进一步调整才能体现其价值和生命力，因此提供的 3ds Max 模型是真正有价值、可随时提取调整用的部分。必须说明的是，书中收录的效果图均为原始模型经过 lightscape 渲染和 photoshop 后处理过的成图，是为读者了解后处理效果提供直观准确的参考，与 3ds Max 直接渲染的效果有一定区别。

著 者
2015 年 2 月

As a well-known decoration company, Guoguangyiye Decoration Group have acquired thousands of international, national and provincial design awards, such as "Top 50 architectural decoration company in China(2012, honored by China Building Decoration Association, CBDA)", "Outstanding Interior Design Companies in China(2011-2012 and 2012-2013, honored by China National Interior Decoration Association, CIDA)", "Top 10 Candlewood Design Companies in China(2012, honor by Chinese and Foreign Hotel Argument)," "The Best Interior Decoration Association of Chinese Home Decoration(2013, honored by CBDA), "Top 10 Interior Design Companies in China (1989~2009)", "Top 10 China Interior Design Institutions (1988~2008)" , "2012 China top 10 Hotel Design Institutions", "China Top 10 Home Decoration Enterprises (1997~2007)" and "Well-Known Brand of Fujian".

In Guoguangyiye Decoration Group, a dozen of architects have be granted as "National Excellent Architect of China", and 79 architects have awarded as "Excellent Architect of Fujian province/Fuzhou", 56 architects have won top prize of national, Fujian provincial or Fuzhou. The chief architect Mr. Bin Ye has wined the award of "Distinguished Achievement Award of Chinese Interior Design", and awarded twice "China Top 10 Interior Design Architect". Mr. Meng Ye was awarded "Outstanding Architect of China (1989~2009)". Naturally these achievements have been accomplished because of the high level interior designs of Guoguangyiye, but obviously cannot be attained without high level professional effect drawing that presents the design intent of architects incisively and vividly. Therefore as a product of the collective efforts of architect and graphic designer, it is closely related to the success of project design.

Since 2004, Guoguangyiye Decoration Group has published 12 series of books on design model database with Fujian Science and Technology Press and all of them have gained wide popularity by their richness and practicality. Therefore, this year we will continue to publish *2015 Interion Design Models Intergration* Space Models Integration. This new series consists of over 1600 chic 3ds Max scenario models of various style interior designs created by Guoguangyiye Decoration Group during 2014~2015. Being a model database, they could also be used as beneficial references for interior design.

The enclosed CD contains original 3ds Max models of decoration effect drawings and all the map files used in order to create them. Due to the continuous upgrading of 3ds Max software, version 2011 was adopted in the drawing of these models which are arranged in the order of the pictures to make them easily accessible. Since as only models that can be further adjusted are valuable, the 3ds Max moulds provided are all of true value and readily available. It should be noted that, all the effect drawings in the books are pictures rendered by lightscape and dealt with by Photoshop, to give an intuitive and precise reference for readers on the after effects which are different from those rendered directly by 3ds Max.

目录 / *CONTENTS*

办公空间
OFFICE SPACE

001
总经理办公室
General Manager's Office

002
会议室
Meeting room

003
办公室
Office

004
多功能厅
Multiple-function hall

005
会议厅
Conference room

006

气象局
Meteorologic Bureau

将乐县气象局

007

走廊
Corridor

008

办公室
Office

009
总经理办公室
General Manager's Office

010
会议室
Meeting room

011
总裁办公室
President's Office

012
卫生间
Washroom

013
卫生间
Washroom

014
办公室
Office

015
办公区
Office area

016

接待室
Reception room

017

总裁办公室
President's Office

018
总裁办公室
President's Office

019
办公区
Office area

020
会议室
Meeting room

021
办公区
Office area

022
电梯厅
Elevator hall

023
电梯厅
Elevator hall

024
门厅
Foyer

025

会议室
Meeting room

026

大堂
Lobby

027
办公区
Office area

028
办公区
Office area

029
餐厅
Restaurant

030
门厅
Foyer

031
接待室
Reception room

032
门厅
Foyer

033
会议室
Meeting room

034
会议厅
Conference room

035
董事长办公室
Chairman's Office

036
董事长办公室
Chairman's Office

037
会议厅
Conference room

038
电梯厅
Elevator hall

039
卫生间
Washroom

040
服务大厅
Service hall

041
总裁办公室
President's Office

042

办公区
Office area

043

办公区
Office area

044

总经理办公室
General Manager's Office

Restaurant

045

多功能厅
Multiple-function hall

046

餐厅
Restaurant

047
总经理办公室
General Manager's Office

048
服务大厅
Service hall

049

电梯厅
Elevator hall

050

洗手间
Washroom

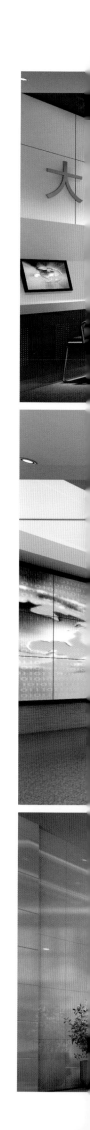

051
办公区
Office area

052
办公区
Office area

053
大堂
Lobby

054
接待室
Reception room

055
会议室
Meeting room

056
办公室
Office

057
接待厅
Reception hall

058
总裁办公室
President's Office

060
办公室
Office

061
走廊
Corridor

059
大堂
Lobby

062
办公区
Office area

063
多功能厅
Multiple-function hall

064
办公区
Office area

065
多功能厅
Multiple-function hall

066
门厅
Foyer

067
门厅
Foyer

068
接待厅
Reception hall

069
办公室
Office

070
餐厅包间
Restaurant room

071
电梯厅
Elevator hall

072
门厅
Foyer

073
接待厅
Reception hall

074
门厅接待区
The lobby reception area

075
多功能厅
Multiple-function hall

076
走廊
Corridor

077
走廊
Corridor

078
主控室
Master control room

079
大厅
Hall

080

外观
Appearance

081

服务厅
Service hall

082

办公区
Office area

083

门厅接待区
The lobby reception area

084

办公室
Office

085

总经理办公室
General Manager's Office

086

多功能厅
Multiple-function hall

087
董事长办公室
Chairman's Office

088
董事长办公室
Chairman's Office

089
总裁办公室
President's Office

090

办公区
Office area

091

门厅
Foyer

092

门厅
Foyer

093

多功能厅
Multiple-function hall

094

总经理办公室
General Manager's Office

095

休息室
Retiring room

096
多功能厅
Multiple-function hall

097
多功能厅
Multiple-function hall

098

营业大厅
Business hall

099
大堂
Lobby

100
大厅
Hall

101
营业大厅
Business hall

102
会议室
Meeting room

103
走廊
Corridor

104
会议室
Meeting room

105
电梯厅
Elevator hall

106
电梯厅
Elevator hall

107
视听室
Audio visual room

中国人民解放军73307部队
文化活动中心

听党指挥 能打胜仗 作风优良

108
大堂
Lobby

109
多功能厅
Multiple-function hall

110
多功能厅
Multiple-function hall

111
开放式办公区
Open office area

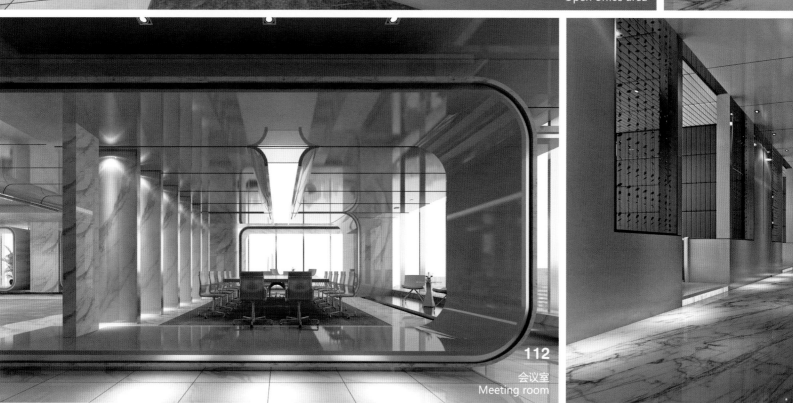

112
会议室
Meeting room

113
前厅
Vestibule

114
贵宾室
VIP room

115
会议室
Meeting room

116
电梯厅
Elevator hall

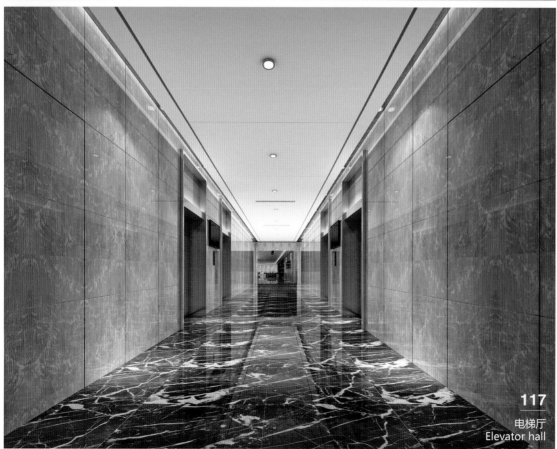

117
电梯厅
Elevator hall

118
会客室
Reception room

119
走廊
Corridor

120
电梯厅
Elevator hall

121
办公区
Office area

122
大堂
Lobby

123
会议室
Meeting room

124
办公区
Office area

125
大堂
Lobby

126
门厅
Foyer

127
大堂
Lobby

128
走廊
Corridor

129
会议室
Meeting room

130
门厅
Foyer

131
会议室
Meeting room

132
监控大厅
Monitor hall

133
餐厅
Restaurant

134
会议室
Meeting room

135
服务大厅
Service hall

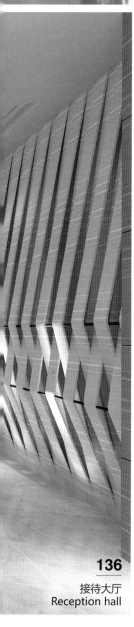

136
接待大厅
Reception hall

137
走廊
Corridor

138
走廊
Corridor

139
办公区
Office area

140

会议室
Meeting room

141

会议室
Meeting room

142
接待室
Reception room

143
会议室
Meeting room

145
办公区
Office area

144
门厅
Foyer

146
服务大厅
Service hall

147

办公区
Office area

148

接待大厅
Reception hall

149
走廊
Corridor

150
服务大厅
Service hall

151
办公区
Office area

152
电梯厅
Elevator hall

153
卫生间
Washroom

154
大堂
Lobby

155
大厅
Hall

157

中庭回廊
Atrium gallery

156

营业大厅
Business hall

158

门厅
Foyer

159

电梯厅
Elevator hall

160

洗手间
Washroom

161
餐厅
Restaurant

162
大厅
Hall

163
会议室
Meeting room

164
会议室
Meeting room

165
接待室
Reception room

166
办公区
Office area

168
服务大厅
Service hall

167
大厅
Hall

169
休息室
Retiring room

170

会议室
Meeting room

171

多功能厅
Multiple-function hall

172
办公区
Office area

173
总经理办公室
General Manager's Office

174
办公室
Office

175
多功能室
Multiple-function room

176
总裁办公室
President's Office

177
餐厅
Restaurant

178
外观
Appearance

179
会议室
Meeting room

并山社区一站式综合服务站

180
服务大厅
Service hall

181
会议室
Meeting room

182
大堂
Lobby

183
接待厅
Reception hall

184
门厅
Foyer

185
大堂
Lobby

186
办公室
Office

187
门厅
Foyer

188
服务大厅
Service hall

189
外观
外观效果图

190
电梯厅
Elevator hall

191

多功能厅
Multiple-function hall

192

餐厅包间
Restaurant room

193
电梯厅
Elevator hall

194
总经理办公室
General Manager's Office

195
洗手间
Washroom

196
电梯厅
Elevator hall

197
服务大厅
Service hall

198
走廊
Corridor

199
大堂
Lobby

200
门厅
Foyer

201
多功能室
Multiple-function room

202
办公室
Office

203
大堂
Lobby

204
董事长办公室
Chairman's Office

205
大堂
Lobby

206
大堂
Lobby

酒店空间
HOTEL SPACE

208
大堂
Lobby

209
接待厅
Reception hall

210
餐厅
Restaurant

211
门厅
Foyer

212
门厅
Foyer

213
电梯厅
Elevator hall

214
走廊
Corridor

216

电梯厅
Elevator hall

217

门厅过道
Hallway

218
标间
Standard room

219
标间
Standard room

220
酒廊
Lounge

222

多功能室
Multiple-function room

221
豪华包间
Deluxe room

223
包间
Compartment

224
标间
Standard room

225
商务标间
Business standard room

226
乒乓球室
Table tennis room

227
餐厅
Restaurant

228
接待厅
Reception hall

229

大床房
Big bed room

230

豪华套房
Deluxe suite

231
门厅过道
Hallway

232
茶室
Tearoom

233

豪华套房
Deluxe suite

234

大床房
Big bed room

235
自助餐厅
Buffet restaurant

236
大堂吧台区
The lobby bar

237
包间
Compartment

238
门厅过道
Hallway

239
商务标间
Business standard room

240
宴会厅
Banquet hall

241
餐厅
Restaurant

242
茶室
Tearoom

商业空间
COMMERCIAL SPACE

243
营业厅
Business hall

244
营业厅
Business hall

245
营业厅
Business hall

购物中心大堂
Shopping Center lobby

247
服装专卖店
Clothing store

248
服装专卖店
Clothing store

249
服装专卖店
Clothing store

250
购物中心大堂
Shopping Center lobby

251
营业厅
Business hall

252

营业厅
Business hall

253

营业厅
Business hall

254
接待区
Reception area

255
专卖店
Speciality stores

256
酒楼门厅
Hotel lobby

257
服装专卖店
Clothing store

258
服装专卖店
Clothing store

259
购物中心
Shopping Center

260
餐厅
Restaurant

261
服装专卖店
Clothing store

262
购物中心
Shopping Center

263
营业厅
Business hall

264
购物中心
Shopping Center

265
银行大厅
Bank hall

266
银行大厅
Bank hall

267
接待区
Reception area

268
购物中心
Shopping Center

269
购物中心
Shopping Center

270
甜品店
Dessert shop

271
商场
market

272
银行大厅
Bank hall

273
餐厅
Restaurant

274
营业厅
Business hall

275
商场
market

276
营业厅
Business hall

购物中心大堂
Shopping Center lobby

279

贵宾室
VIP room

278

银行大厅
Bank hall

280

贵宾室
VIP room

281
茶室
Tearoom

282
展览室
Exhibition room

283
超市
Supermarket

284
餐厅
Restaurant

285
银行大厅
Bank hall

286
营业厅
Business hall

287
银行大厅
Bank hall

288
营业厅
Business hall

289
营业厅
Business hall

290
超市
supermarket

291
银行大厅
Bank hall

292
营业厅
Business hall

293
酒楼门厅
Hotel lobby

294
餐馆门厅
Restaurant lobby

295
咖啡店
Coffee shop

296
咖啡店
Coffee shop

297
咖啡店
Coffee shop

298
咖啡店
Coffee shop

299
咖啡店
Coffee shop

300
咖啡店
Coffee shop

301
咖啡店
Coffee shop

302
咖啡店
Coffee shop

303
咖啡店
Coffee shop

房产空间
ESTATE SPACE

306
售楼大厅
Sales hall

3
售楼大厅
Sales hall

305
售楼大厅
Sales hall

307
售楼办
Sales Office

308
入户门厅
Entrance hall

309
入户电梯厅
Household elevator hall

310

售楼大厅
Sales hall

311

接待室
Reception room

312
售楼大厅
Sales hall

313
售楼大厅
Sales hall

314
样板房
Show flat

315
售楼厅洽谈区
Sales hall discussion area

316
售楼厅水吧区
Sales hall water area

317
会议室
Meeting room

318
洗手间
Washroom

319
洗手间
Washroom

320

售楼大厅
Sales hall

321

走廊
Corridor

322
样板房
Show flat

323
健身房
Fitness centre

324
电梯大堂
Elevator lobby

325
样板房
Show flat

326
售楼大厅
Sales hall

327
会议室
Meeting room

328
游泳池
Swimming pool

329
样板房
Show flat

330
售楼厅游乐区
Sales hall recreation area

331
售楼大厅
Sales hall

332

地下停车场
Underground parking garage

333

售楼大厅
Sales hall

334

外观
Appearance

835
接待区
Reception area

336
办公区
Office area

337

大堂电梯厅
Lobby elevator hall

338

大堂电梯厅
Lobby elevator hall

339
样板房
Show flat

340
洗手间
Washroom

341
洗手间
Washroom

343
影音室
Video room

344
样板房
Show flat

345
样板房
Show flat

346
样板房
Show flat

347
入户大堂
Entrance hall

348
售楼厅洽谈区
Sales hall discussion area

349
售楼厅洽谈区
Sales hall discussion area

350
样板房
Show flat

351
样板房
Show flat

352
售楼大厅
Sales hall

353
门厅
Foyer

354

样板房
Show flat

355

电梯厅
Elevator hall

356

走廊
Corridor

357

售楼大厅
Sales hall

358

大堂电梯厅
Lobby elevator hall

359

样板房
Show flat

360

样板房
Show flat

362
样板房
Show flat

363
样板房
Show flat

364
样板房
Show flat

365
样板房
Show flat

366
样板房
Show flat

367
售楼厅洽谈区
Sales hall discussion area

368
样板房
Show flat

369
样板房
Show flat

370
大堂走道
Lobby walkway

371
大堂电梯厅
Lobby elevator hall

372
门厅
Foyer

373
沙盘区
Sand table area

374
样板房
Show flat

图书在版编目（CIP）数据

2015 室内设计模型集成 . 公共空间 / 叶斌 , 叶猛
著 .—福州 : 福建科学技术出版社 , 2015.4
　ISBN 978-7-5335-4738-7

　Ⅰ . ① 2… Ⅱ . ① 叶… ② 叶… Ⅲ . ① 公共建筑－室内
装饰设计－作品集－中国－现代 Ⅳ . ① TU24

　　中国版本图书馆 CIP 数据核字 (2015) 第 044421 号

书　　名　2015 室内设计模型集成　公共空间
著　　者　叶斌　叶猛
出版发行　海峡出版发行集团
　　　　　福建科学技术出版社
社　　址　福州市东水路 76 号（邮编 350001）
网　　址　www.fjstp.com
经　　销　福建新华发行（集团）有限责任公司
印　　刷　恒美印务（广州）有限公司
开　　本　635 毫米 ×965 毫米　1/8
印　　张　22
图　　文　176 码
版　　次　2015 年 4 月第 1 版
印　　次　2015 年 4 月第 1 次印刷
书　　号　ISBN 978-7-5335-4738-7
定　　价　248.00 元
　　书中如有印装质量问题，可直接向本社调换